Division

H.S. Lawrence

Illustration by
Kathy Kifer and Dahna Solar

Special thanks to:
Holly Dye, Derrick Hoffman, Jane Tory, Carrie Hernandez,
Caroline Jeanbart, Susan Rovetta, and Cecily Cleveland

Published by
Garlic Press
100 Hillview Lane #2
Eugene, OR 97401

ISBN 0-931993-59-8
Order Number GP-059

Overview: Math and Animal Science

The Puzzles and Practice Series builds basic **math skills** and acquaints students with **animal science**. The Series is also designed to challenge skills associated with following directions, simple logic, visual discrimination (all puzzle assembly skills), and motor skills (cutting and pasting).

Practice Pages illustrate math skills step-by-step, then provide extended practice. **Puzzle Pages** contain twelve-piece puzzles that when assembled reveal a fascinating animal. This book in the Series features Grassland Animals.

Grassland Animal Reference Cards, found on the last three pages of this book, provide further information for students. In addition, for parents and teachers, the inside front cover provides **background information** on Grassland Animals.

Helping Teachers and Parents

There are two pages for each of the twelve lessons- a Practice Page and a Puzzle Page. Each page can be used independently; however, the Puzzles and Practice Series has incorporated a special feature that encourages the use of both pages at one time.

Special Feature - If you hold a *Puzzle Page* up to the light, you will see the same problems showing in the center of the puzzle pieces (actually showing through from the *Practice Page*) that are to the left of the puzzle pieces on the Puzzle Page. This feature is useful so a student will not lose the potential for the answer after he or she has cut out the puzzle piece. This feature is also useful if a student does not follow directions and cuts out all puzzle pieces at one time.

Table of Contents

NAME
NOMBRE ______________________

$$1\overset{5}{\overline{)5}} \qquad\qquad 1\overset{8}{\overline{)8}}$$

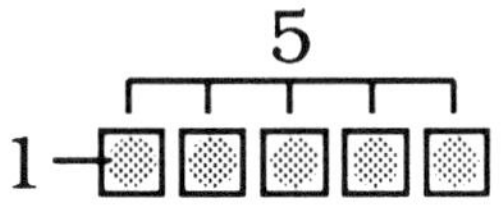

$\begin{array}{r} 5 \\ \times\ 1 \\ \hline 5 \end{array}$

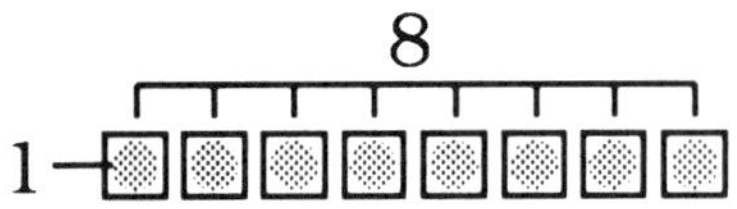

$\begin{array}{r} 8 \\ \times\ 1 \\ \hline 8 \end{array}$

$1\overline{)4}$	$1\overline{)7}$	$1\overline{)10}$	$1\overline{)2}$	$1\overline{)11}$	$1\overline{)0}$	$1\overline{)1}$	$1\overline{)8}$
$1\overline{)10}$	$1\overline{)3}$	$1\overline{)5}$	$1\overline{)0}$	$1\overline{)9}$	$1\overline{)12}$	$1\overline{)6}$	$1\overline{)9}$
$1\overline{)2}$	$1\overline{)5}$	$1\overline{)8}$	$1\overline{)6}$	$1\overline{)12}$	$1\overline{)1}$	$1\overline{)4}$	$1\overline{)10}$
$1\overline{)3}$	$1\overline{)9}$	$1\overline{)11}$	$1\overline{)4}$	$1\overline{)7}$	$1\overline{)12}$	$1\overline{)1}$	$1\overline{)8}$

NAME
NOMBRE ______________________

Instructions:

1. **Answer all the math problems first.**
2. **Cut out one puzzle piece at a time.**
3. **Paste the puzzle piece in the box with the same answer.**

Instrucciones:

1. Conteste todos los problemas de matemáticas primero.
2. Recorte una pieza del rompecabezas a la vez.
3. Pegue la pieza del rompecabezas en el recuadro que tiene la misma respuesta.

5	**2**	**10**	**7**
8	**6**	**12**	**3**
4	**11**	**1**	**9**

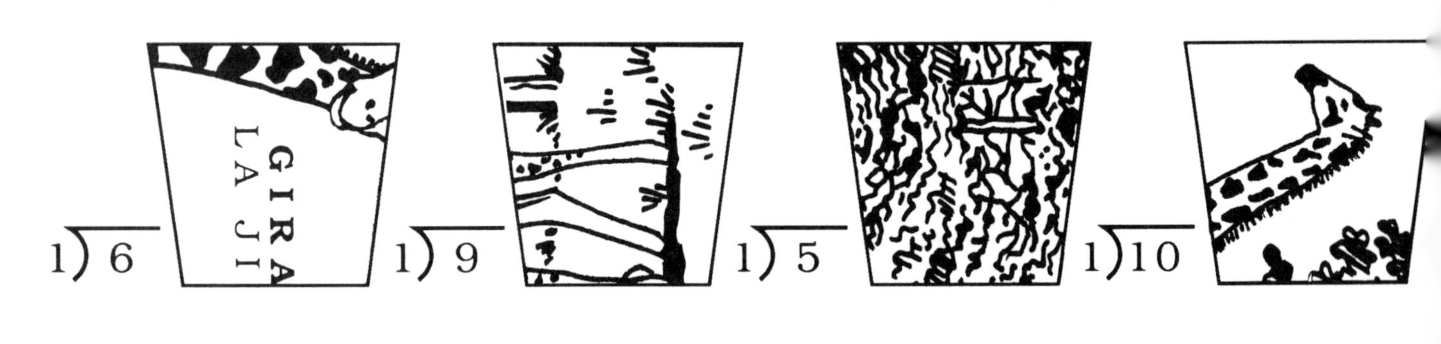

NAME
NOMBRE ______________________________

$$2\overline{)6}\quad 3$$

$$2\overline{)8}\quad 4$$

$3 \times 2 = 6$

$4 \times 2 = 8$

$2\overline{)20}$	$2\overline{)14}$	$2\overline{)2}$	$2\overline{)22}$	$2\overline{)6}$	$2\overline{)12}$	$2\overline{)24}$	$2\overline{)4}$
$2\overline{)8}$	$2\overline{)4}$	$2\overline{)16}$	$2\overline{)10}$	$2\overline{)2}$	$2\overline{)18}$	$2\overline{)10}$	$2\overline{)6}$
$2\overline{)4}$	$2\overline{)8}$	$2\overline{)24}$	$2\overline{)0}$	$2\overline{)18}$	$2\overline{)22}$	$2\overline{)6}$	$2\overline{)20}$
$2\overline{)12}$	$2\overline{)16}$	$2\overline{)22}$	$2\overline{)8}$	$2\overline{)14}$	$2\overline{)2}$	$2\overline{)20}$	$2\overline{)0}$

Lesson
Lección 2

NAME
NOMBRE ______________________

Instructions:

1. **Answer <u>all</u> the math problems first.**
2. **Cut out <u>one</u> puzzle piece at a time.**
3. **Paste the puzzle piece in the box with the same answer.**

Instrucciones:

1. Conteste todos los problemas de matemáticas primero.
2. Recorte una pieza del rompecabezas a la vez.
3. Pegue la pieza del rompecabezas en el recuadro que tiene la misma respuesta.

7	**1**	**6**	**4**
12	**3**	**10**	**5**
9	**8**	**2**	**11**

$2\overline{)10}$ $2\overline{)2}$ $2\overline{)16}$ $2\overline{)8}$

$2\overline{)6}$ $2\overline{)18}$ $2\overline{)24}$ $2\overline{)4}$

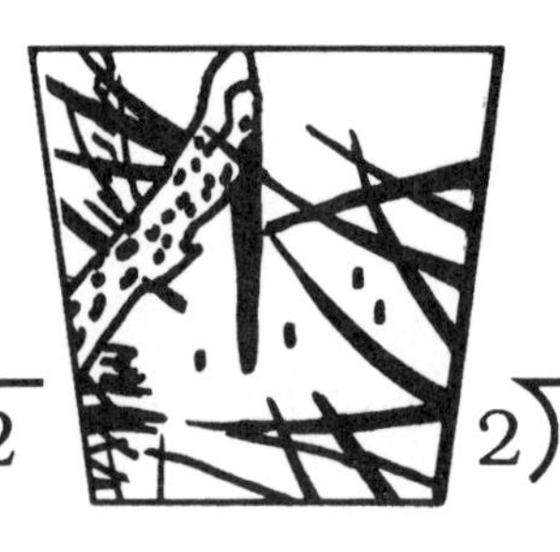

$2\overline{)20}$ $2\overline{)14}$ $2\overline{)22}$ $2\overline{)12}$

NAME
NOMBRE ______________________

$$3\overline{)\ 9}^{\,3}$$

$$3\overline{)12}^{\,4}$$

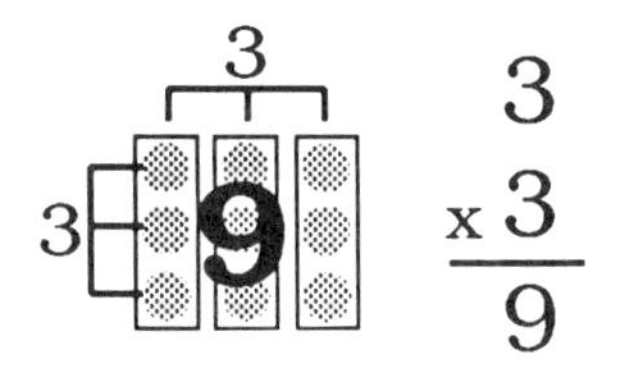

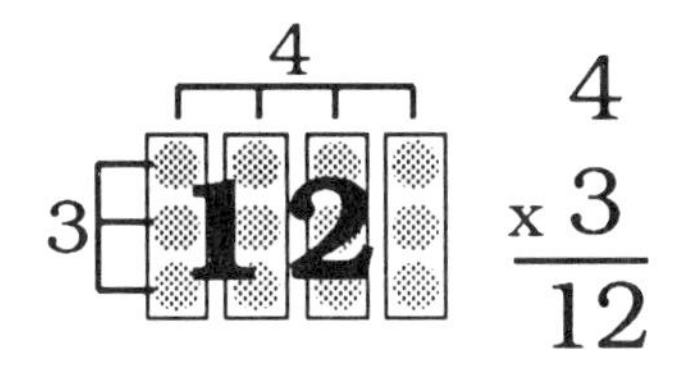

$3\overline{)\ 9}$ $3\overline{)\ 6}$ $3\overline{)24}$ $3\overline{)18}$ $3\overline{)\ 0}$ $3\overline{)36}$ $3\overline{)27}$ $3\overline{)12}$

$3\overline{)27}$ $3\overline{)15}$ $3\overline{)\ 3}$ $3\overline{)21}$ $3\overline{)30}$ $3\overline{)\ 3}$ $3\overline{)15}$ $3\overline{)33}$

$3\overline{)36}$ $3\overline{)\ 6}$ $3\overline{)18}$ $3\overline{)\ 0}$ $3\overline{)12}$ $3\overline{)21}$ $3\overline{)33}$ $3\overline{)18}$

$3\overline{)24}$ $3\overline{)30}$ $3\overline{)\ 9}$ $3\overline{)36}$ $3\overline{)21}$ $3\overline{)\ 3}$ $3\overline{)\ 6}$ $3\overline{)15}$

Lesson 3
Lección 3

NAME
NOMBRE ______________________________

Instructions:

1. **Answer <u>all</u> the math problems first.**
2. **Cut out <u>one</u> puzzle piece at a time.**
3. **Paste the puzzle piece in the box with the same answer.**

Instrucciones:

1. Conteste <u>todos</u> los problemas de matemáticas primero.
2. Recorte <u>una</u> pieza del rompecabezas a la vez.
3. Pegue la pieza del rompecabezas en el recuadro que tiene la misma respuesta.

2	11	6	5
10	3	1	8
7	9	12	4

$3\overline{)15}$ $3\overline{)30}$ $3\overline{)3}$ $3\overline{)27}$

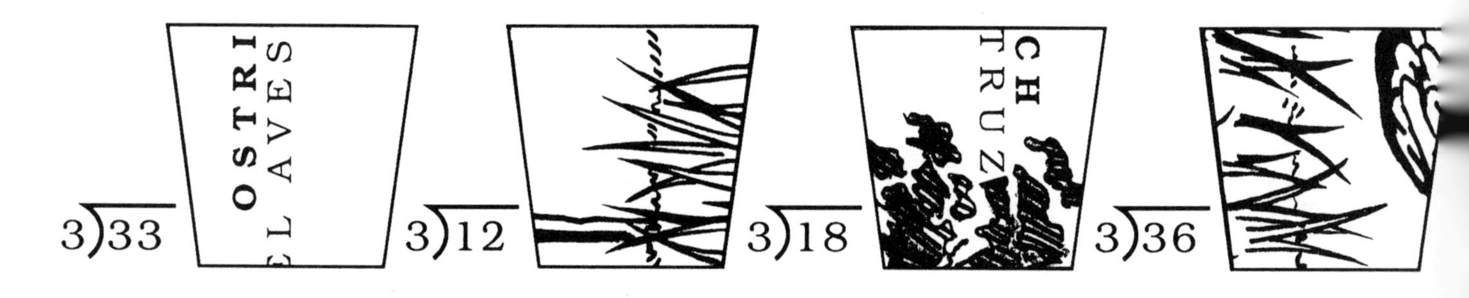

$3\overline{)33}$ $3\overline{)12}$ $3\overline{)18}$ $3\overline{)36}$

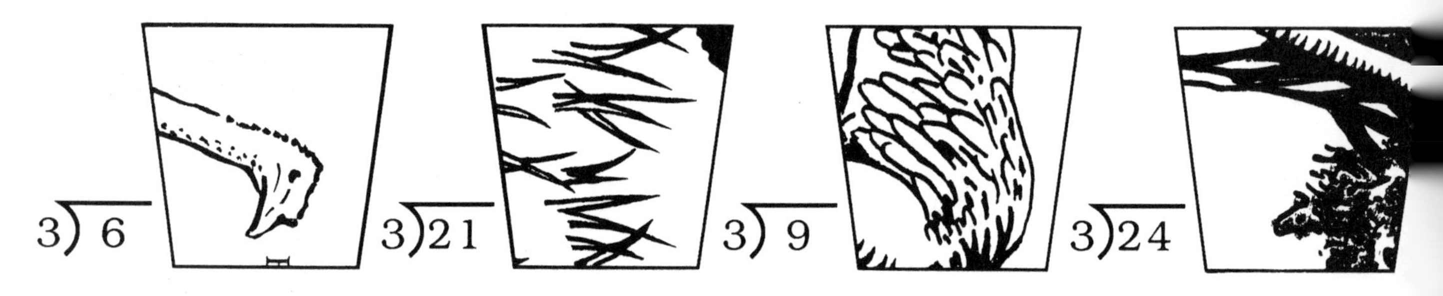

$3\overline{)6}$ $3\overline{)21}$ $3\overline{)9}$ $3\overline{)24}$

NAME
NOMBRE ______________________

$$\begin{array}{r} 6 \\ 4\overline{)24} \end{array} \qquad \begin{array}{r} 4 \\ 4\overline{)16} \end{array}$$

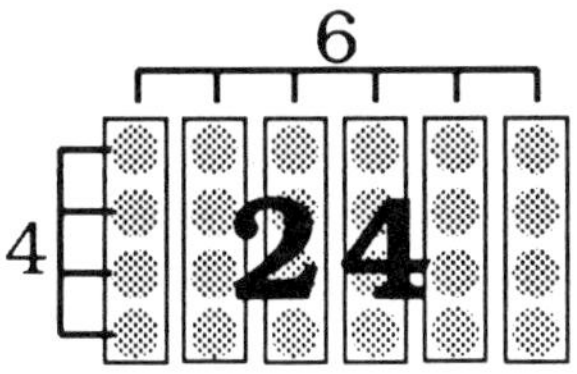

$$\begin{array}{r} 6 \\ \times\ 4 \\ \hline 24 \end{array}$$

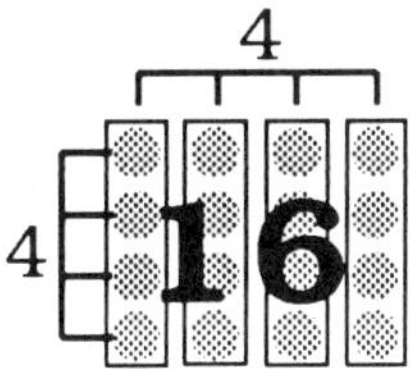

$$\begin{array}{r} 4 \\ \times\ 4 \\ \hline 16 \end{array}$$

$4\overline{)36}$	$4\overline{)12}$	$4\overline{)0}$	$4\overline{)48}$	$4\overline{)8}$	$4\overline{)24}$	$4\overline{)44}$	$4\overline{)16}$
$4\overline{)4}$	$4\overline{)40}$	$4\overline{)28}$	$4\overline{)16}$	$4\overline{)40}$	$4\overline{)28}$	$4\overline{)32}$	$4\overline{)20}$
$4\overline{)8}$	$4\overline{)32}$	$4\overline{)20}$	$4\overline{)48}$	$4\overline{)16}$	$4\overline{)12}$	$4\overline{)44}$	$4\overline{)0}$
$4\overline{)48}$	$4\overline{)4}$	$4\overline{)24}$	$4\overline{)44}$	$4\overline{)36}$	$4\overline{)28}$	$4\overline{)12}$	$4\overline{)32}$

NAME
NOMBRE ____________________

Instructions:

1. **Answer all the math problems first.**
2. **Cut out one puzzle piece at a time.**
3. **Paste the puzzle piece in the box with the same answer.**

Instrucciones:

1. Conteste todos los problemas de matemáticas primero.
2. Recorte una pieza del rompecabezas a la vez.
3. Pegue la pieza del rompecabezas en el recuadro que tiene la misma respuesta.

2	7	9	4
11	5	12	10
3	8	6	1

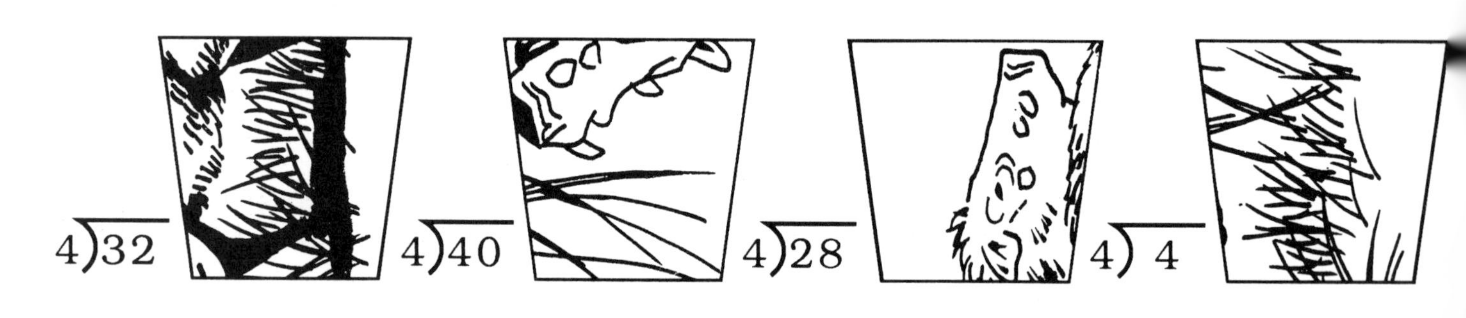

$4\overline{)32}$ $4\overline{)40}$ $4\overline{)28}$ $4\overline{)4}$

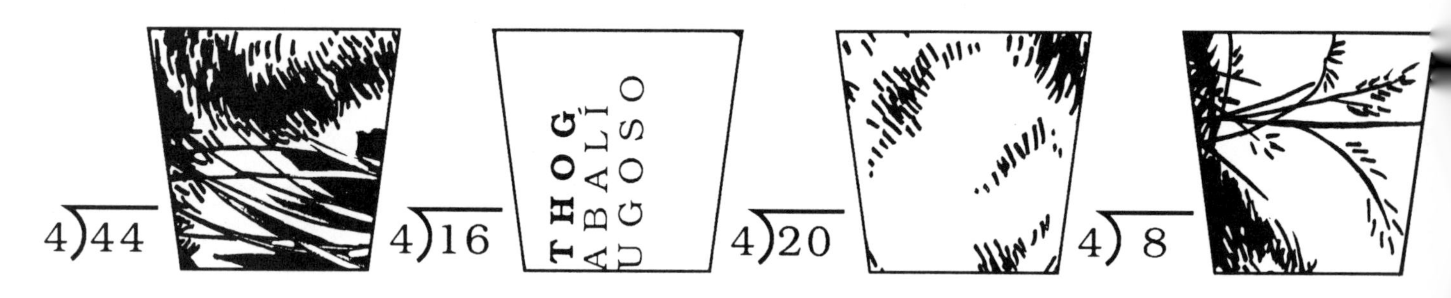

$4\overline{)44}$ $4\overline{)16}$ $4\overline{)20}$ $4\overline{)8}$

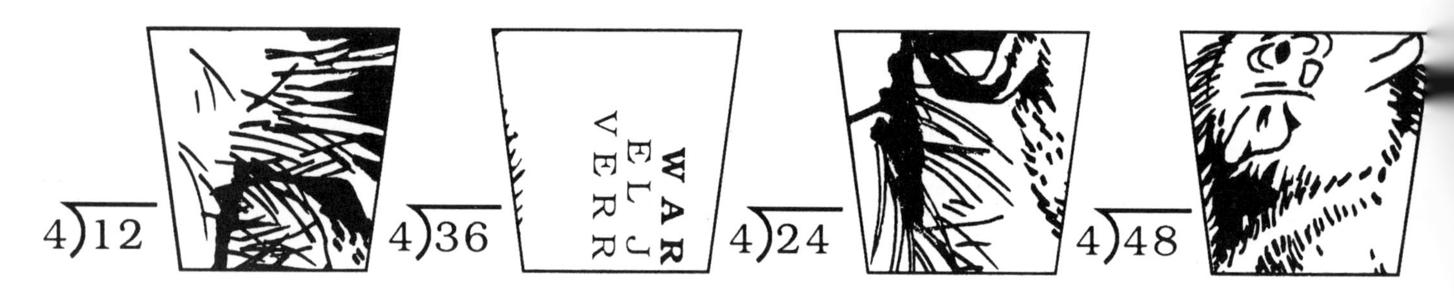

$4\overline{)12}$ $4\overline{)36}$ $4\overline{)24}$ $4\overline{)48}$

NAME
NOMBRE ______________________

$$5\overline{)20}\quad \text{quotient } 4$$

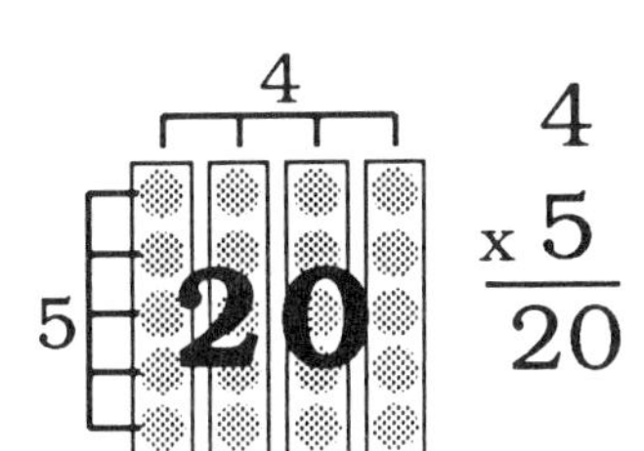

$$\begin{array}{r} 4 \\ \times 5 \\ \hline 20 \end{array}$$

$5\overline{)60}$ $5\overline{)25}$ $5\overline{)45}$ $5\overline{)\ 0}$ $5\overline{)35}$ $5\overline{)20}$ $5\overline{)40}$ $5\overline{)15}$

$5\overline{)45}$ $5\overline{)30}$ $5\overline{)10}$ $5\overline{)50}$ $5\overline{)\ 5}$ $5\overline{)25}$ $5\overline{)55}$ $5\overline{)10}$

$5\overline{)30}$ $5\overline{)\ 5}$ $5\overline{)60}$ $5\overline{)\ 0}$ $5\overline{)15}$ $5\overline{)55}$ $5\overline{)25}$ $5\overline{)50}$

$5\overline{)40}$ $5\overline{)15}$ $5\overline{)50}$ $5\overline{)45}$ $5\overline{)20}$ $5\overline{)60}$ $5\overline{)35}$ $5\overline{)\ 5}$

Lesson 5
Lección 5

NAME
NOMBRE ______________________

Instructions:

1. **Answer all the math problems first.**
2. **Cut out one puzzle piece at a time.**
3. **Paste the puzzle piece in the box with the same answer.**

Instrucciones:

1. Conteste todos los problemas de matemáticas primero.
2. Recorte una pieza del rompecabezas a la vez.
3. Pegue la pieza del rompecabezas en el recuadro que tiene la misma respuesta.

7	**10**	**1**	**5**
4	**2**	**8**	**12**
11	**6**	**3**	**9**

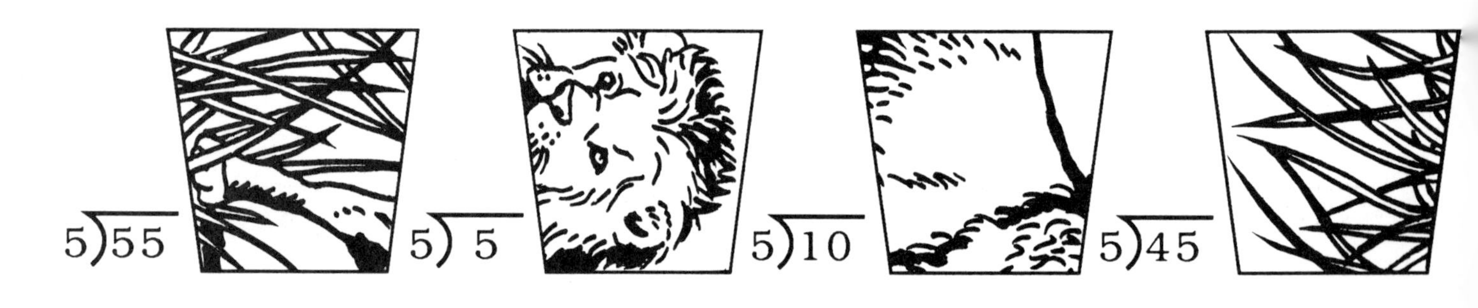

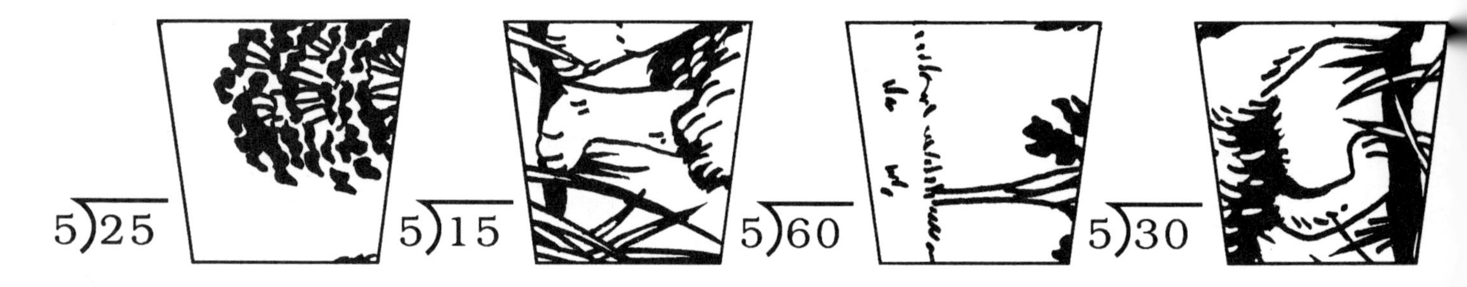

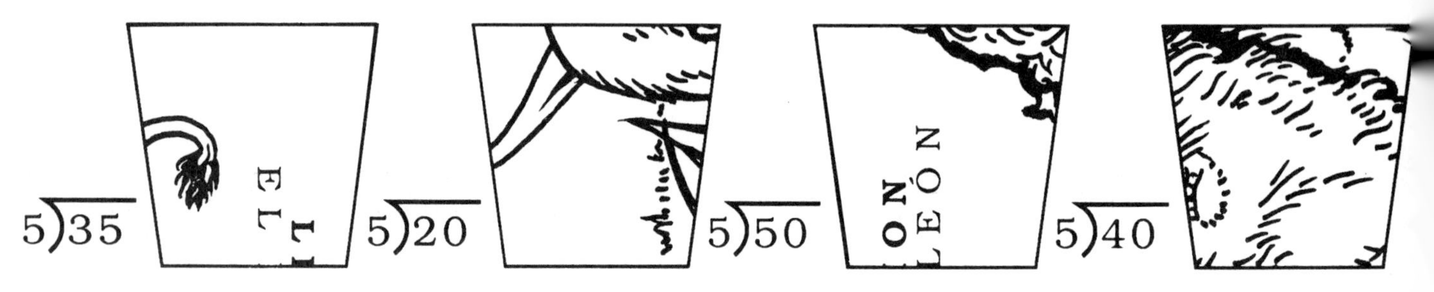

Lesson Lección 6

NAME
NOMBRE ______________________________

$$6\overline{)18}\ \text{with quotient } 3$$

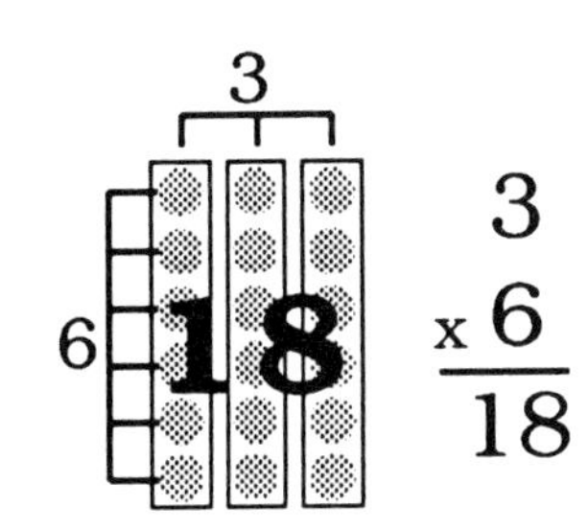

$6\overline{)24}$	$6\overline{)6}$	$6\overline{)48}$	$6\overline{)72}$	$6\overline{)18}$	$6\overline{)0}$	$6\overline{)30}$	$6\overline{)54}$
$6\overline{)30}$	$6\overline{)48}$	$6\overline{)12}$	$6\overline{)42}$	$6\overline{)66}$	$6\overline{)24}$	$6\overline{)42}$	$6\overline{)6}$
$6\overline{)60}$	$6\overline{)36}$	$6\overline{)48}$	$6\overline{)66}$	$6\overline{)6}$	$6\overline{)18}$	$6\overline{)54}$	$6\overline{)72}$
$6\overline{)18}$	$6\overline{)60}$	$6\overline{)72}$	$6\overline{)54}$	$6\overline{)24}$	$6\overline{)66}$	$6\overline{)36}$	$6\overline{)0}$

Lesson 6
Lección 6

NAME
NOMBRE ______________________

Instructions:

1. **Answer <u>all</u> the math problems first.**
2. **Cut out <u>one</u> puzzle piece at a time.**
3. **Paste the puzzle piece in the box with the same answer.**

Instrucciones:

1. Conteste <u>todos</u> los problemas de matemáticas primero.
2. Recorte <u>una</u> pieza del rompecabezas a la vez.
3. Pegue la pieza del rompecabezas en el recuadro que tiene la misma respuesta.

7	2	12	9
4	8	5	3
10	1	11	6

$6\overline{)42}$ $6\overline{)66}$ $6\overline{)12}$ $6\overline{)30}$

$6\overline{)54}$ $6\overline{)6}$ $6\overline{)48}$ $6\overline{)60}$

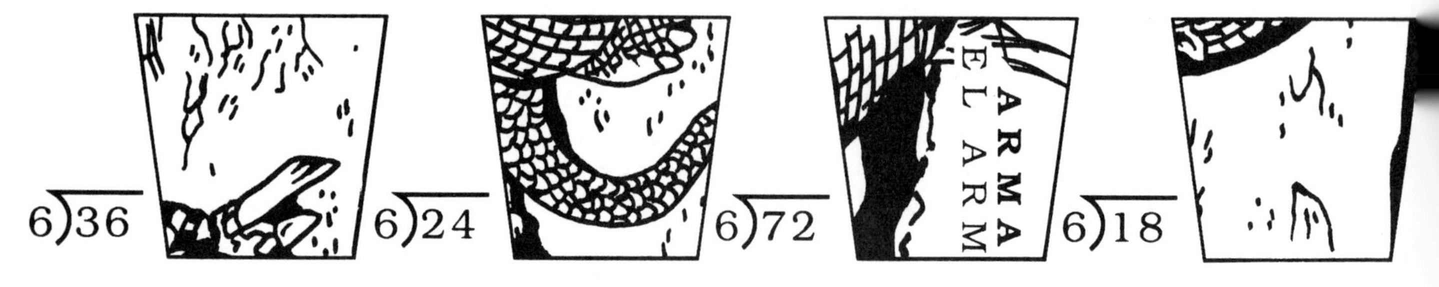

$6\overline{)36}$ $6\overline{)24}$ $6\overline{)72}$ $6\overline{)18}$

NAME
NOMBRE ____________________

$$7\overline{)56}^{\;8}$$

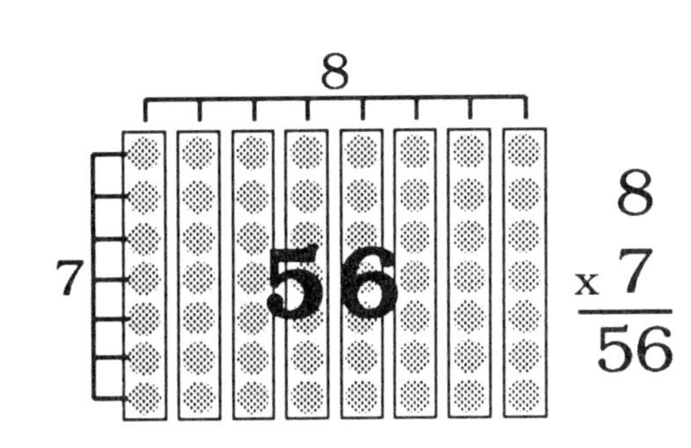

$7\overline{)63}$ $7\overline{)21}$ $7\overline{)84}$ $7\overline{)\ 0}$ $7\overline{)70}$ $7\overline{)49}$ $7\overline{)35}$ $7\overline{)14}$

$7\overline{)28}$ $7\overline{)42}$ $7\overline{)77}$ $7\overline{)21}$ $7\overline{)\ 7}$ $7\overline{)84}$ $7\overline{)56}$ $7\overline{)28}$

$7\overline{)42}$ $7\overline{)14}$ $7\overline{)84}$ $7\overline{)56}$ $7\overline{)35}$ $7\overline{)49}$ $7\overline{)70}$ $7\overline{)\ 0}$

$7\overline{)49}$ $7\overline{)28}$ $7\overline{)63}$ $7\overline{)77}$ $7\overline{)14}$ $7\overline{)42}$ $7\overline{)21}$ $7\overline{)\ 7}$

Lesson 7
Lección 7

NAME
NOMBRE ______________________

Instructions:

1. **Answer all the math problems first.**
2. **Cut out one puzzle piece at a time.**
3. **Paste the puzzle piece in the box with the same answer.**

Instrucciones:

1. Conteste todos los problemas de matemáticas primero.
2. Recorte una pieza del rompecabezas a la vez.
3. Pegue la pieza del rompecabezas en el recuadro que tiene la misma respuesta.

6	**11**	**5**	**9**
7	**3**	**8**	**4**
1	**12**	**2**	**10**

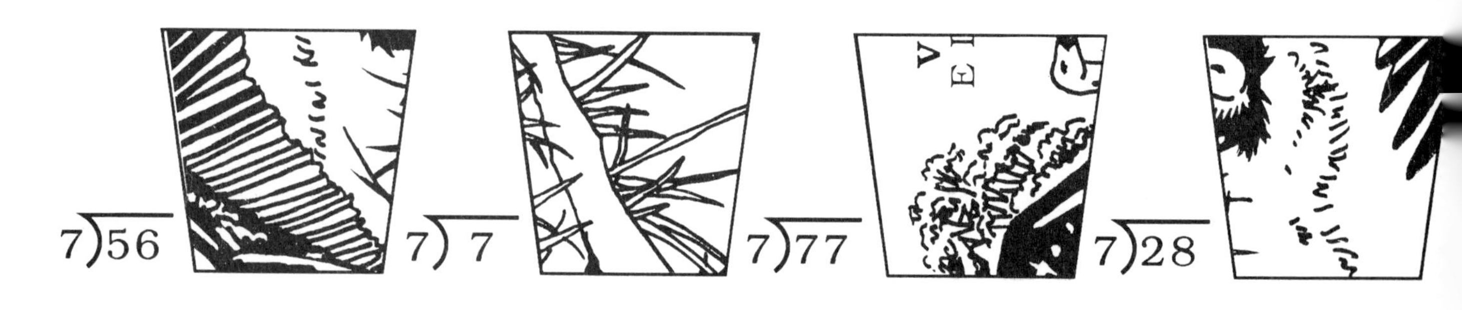

NAME
NOMBRE ______________________________

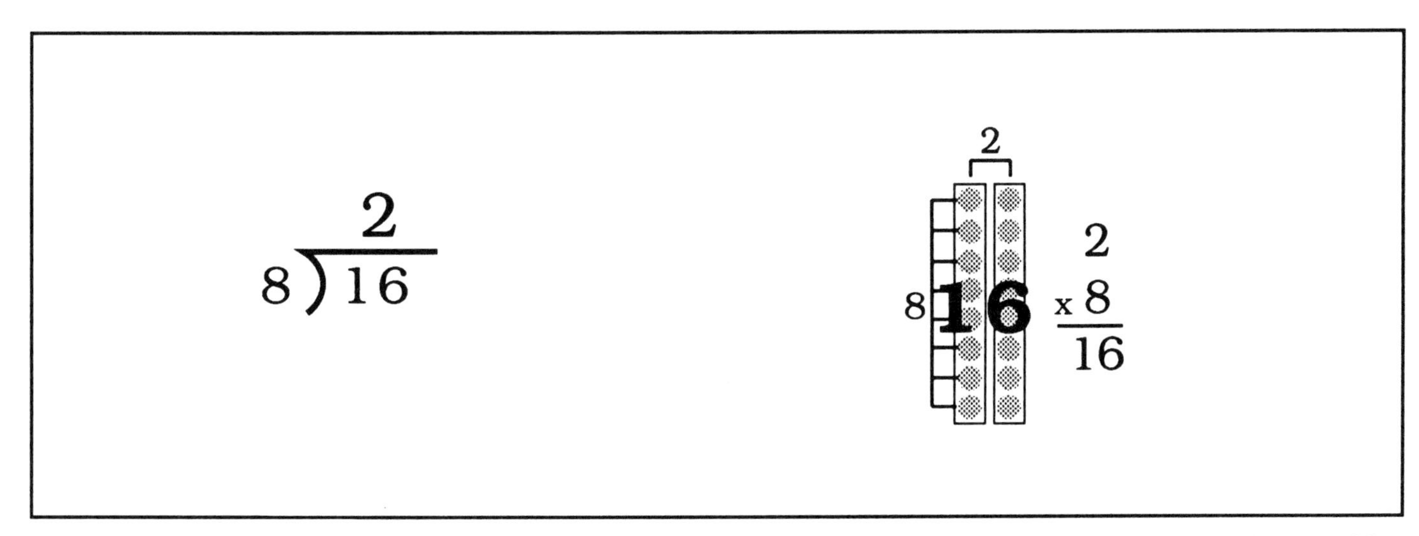

$8\overline{)48}$ $8\overline{)\ 0}$ $8\overline{)16}$ $8\overline{)64}$ $8\overline{)96}$ $8\overline{)32}$ $8\overline{)80}$ $8\overline{)\ 8}$

$8\overline{)24}$ $8\overline{)56}$ $8\overline{)88}$ $8\overline{)\ 0}$ $8\overline{)40}$ $8\overline{)72}$ $8\overline{)16}$ $8\overline{)96}$

$8\overline{)\ 8}$ $8\overline{)40}$ $8\overline{)96}$ $8\overline{)64}$ $8\overline{)32}$ $8\overline{)56}$ $8\overline{)80}$ $8\overline{)24}$

$8\overline{)56}$ $8\overline{)88}$ $8\overline{)72}$ $8\overline{)\ 8}$ $8\overline{)64}$ $8\overline{)24}$ $8\overline{)48}$ $8\overline{)80}$

Lesson
Lección 8

NAME
NOMBRE ______________________

Instructions:

1. **Answer all the math problems first.**
2. **Cut out one puzzle piece at a time.**
3. **Paste the puzzle piece in the box with the same answer.**

Instrucciones:

1. Conteste todos los problemas de matemáticas primero.
2. Recorte una pieza del rompecabezas a la vez.
3. Pegue la pieza del rompecabezas en el recuadro que tiene la misma respuesta.

3	**9**	**1**	**5**
12	**6**	**10**	**8**
11	**7**	**2**	**4**

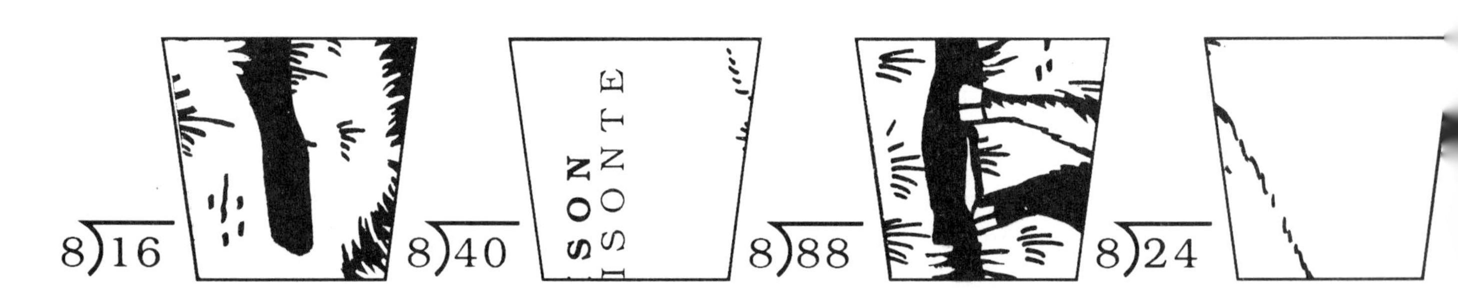

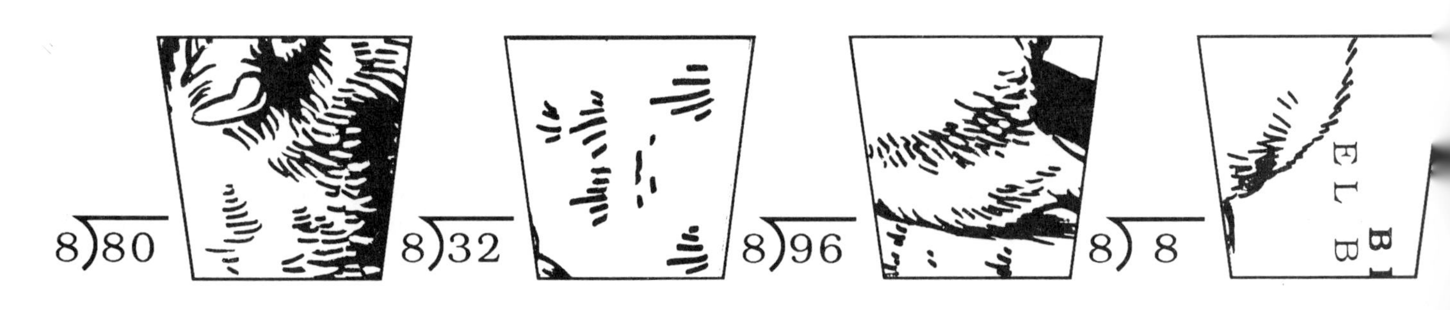

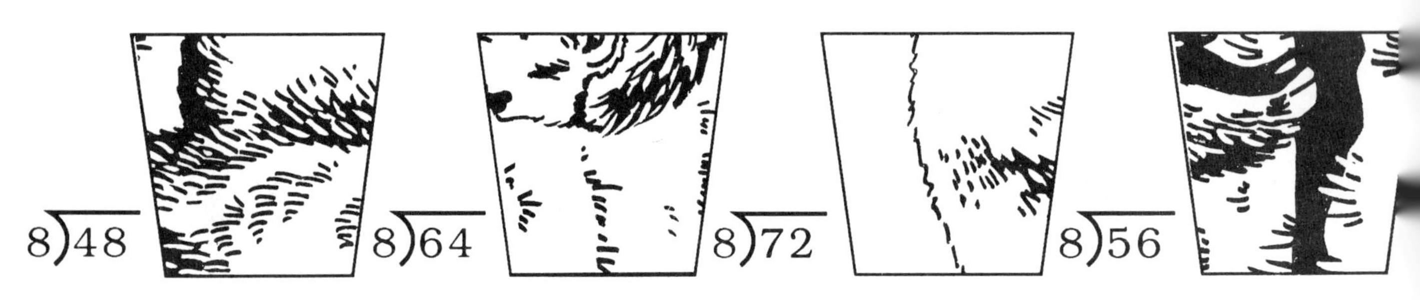

NAME
NOMBRE ______________________

$$9\overline{)63}\quad 7$$

7

9 **63**

$$\begin{array}{r} 7 \\ \times 9 \\ \hline 63 \end{array}$$

$9\overline{)54}$	$9\overline{)27}$	$9\overline{)108}$	$9\overline{)72}$	$9\overline{)90}$	$9\overline{)0}$	$9\overline{)45}$	$9\overline{)18}$
$9\overline{)81}$	$9\overline{)63}$	$9\overline{)9}$	$9\overline{)54}$	$9\overline{)36}$	$9\overline{)72}$	$9\overline{)90}$	$9\overline{)27}$
$9\overline{)72}$	$9\overline{)36}$	$9\overline{)108}$	$9\overline{)18}$	$9\overline{)45}$	$9\overline{)0}$	$9\overline{)27}$	$9\overline{)99}$
$9\overline{)63}$	$9\overline{)81}$	$9\overline{)99}$	$9\overline{)9}$	$9\overline{)54}$	$9\overline{)108}$	$9\overline{)18}$	$9\overline{)63}$

NAME
NOMBRE ______________________________

Instructions:

1. **Answer all the math problems first.**
2. **Cut out one puzzle piece at a time.**
3. **Paste the puzzle piece in the box with the same answer.**

Instrucciones:

1. Conteste todos los problemas de matemáticas primero.
2. Recorte una pieza del rompecabezas a la vez.
3. Pegue la pieza del rompecabezas en el recuadro que tiene la misma respuesta.

5	2	9	4
6	10	12	7
1	11	3	8

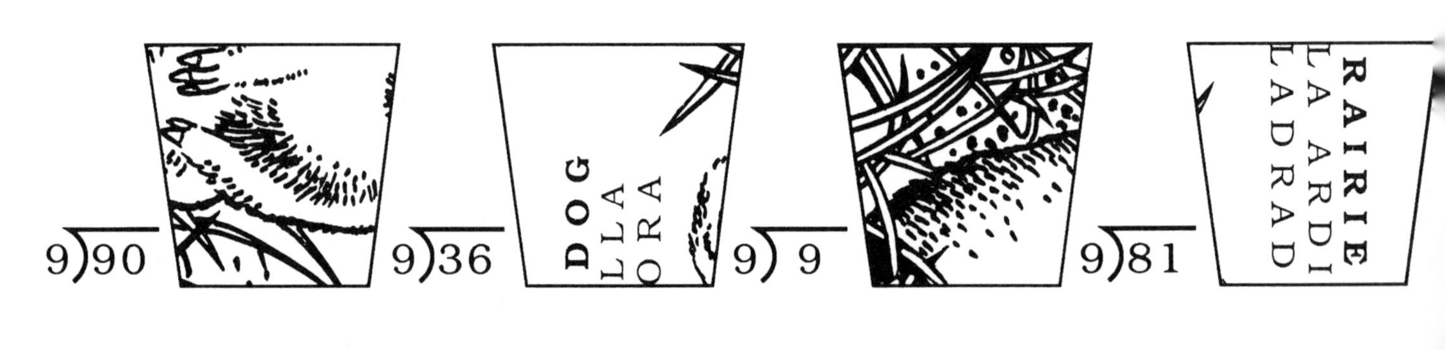

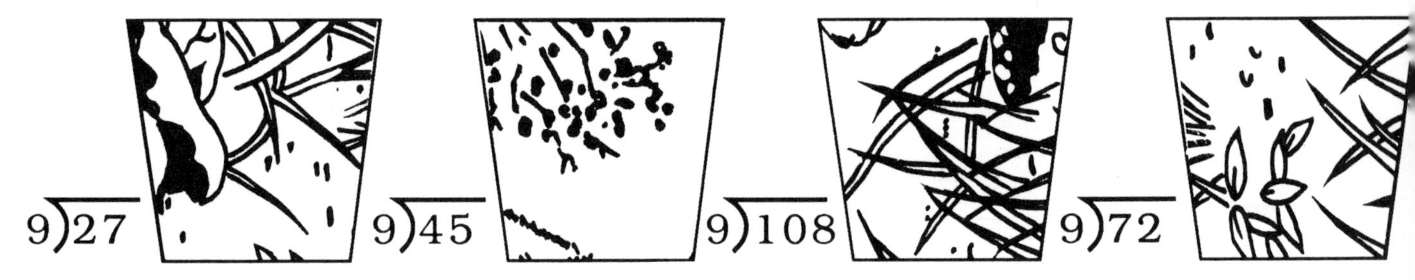

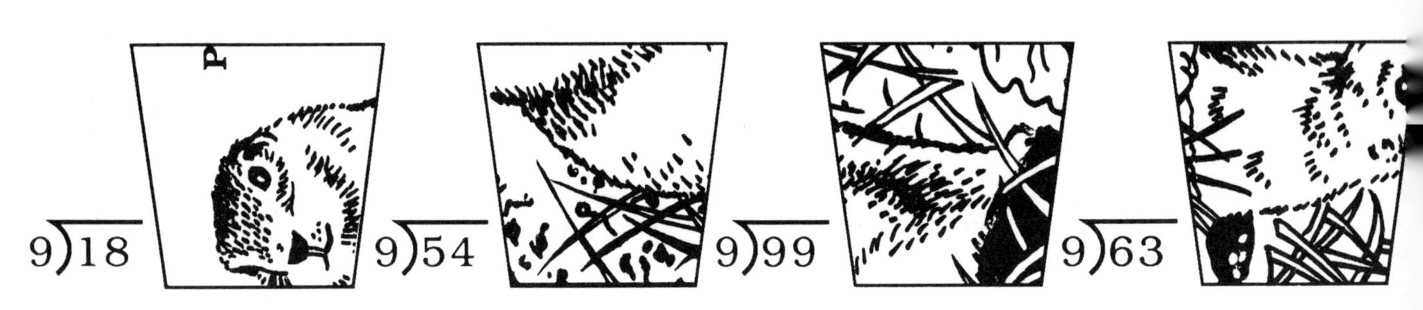

NAME
NOMBRE ______________________________

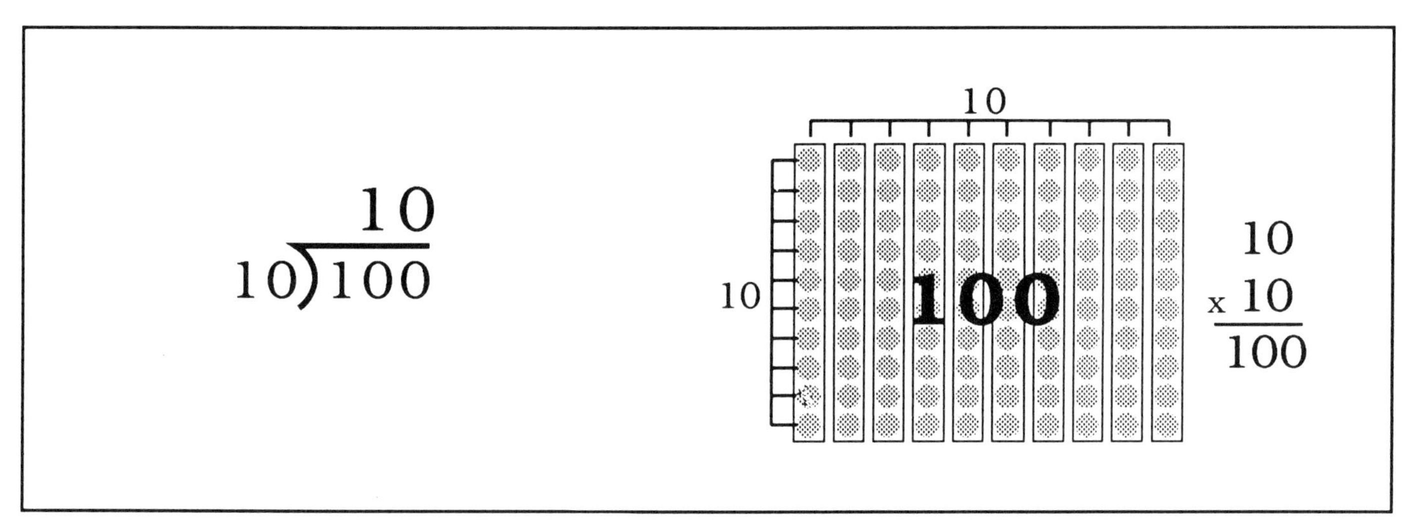

$10\overline{)10}$ $10\overline{)120}$ $10\overline{)50}$ $10\overline{)0}$ $10\overline{)90}$ $10\overline{)20}$ $10\overline{)40}$ $10\overline{)100}$

$10\overline{)30}$ $10\overline{)110}$ $10\overline{)90}$ $10\overline{)10}$ $10\overline{)50}$ $10\overline{)120}$ $10\overline{)70}$ $10\overline{)30}$

$10\overline{)20}$ $10\overline{)100}$ $10\overline{)80}$ $10\overline{)50}$ $10\overline{)120}$ $10\overline{)0}$ $10\overline{)60}$ $10\overline{)90}$

$10\overline{)40}$ $10\overline{)60}$ $10\overline{)100}$ $10\overline{)80}$ $10\overline{)110}$ $10\overline{)30}$ $10\overline{)10}$ $10\overline{)70}$

Lesson / Lección 10

NAME
NOMBRE ______________________________

Instructions:

1. **Answer <u>all</u> the math problems first.**
2. **Cut out <u>one</u> puzzle piece at a time.**
3. **Paste the puzzle piece in the box with the same answer.**

Instrucciones:

1. Conteste <u>todos</u> los problemas de matemáticas primero.
2. Recorte <u>una</u> pieza del rompecabezas a la vez.
3. Pegue la pieza del rompecabezas en el recuadro que tiene la misma respuesta.

3	**8**	**4**	**7**
10	**5**	**11**	**6**
12	**1**	**9**	**2**

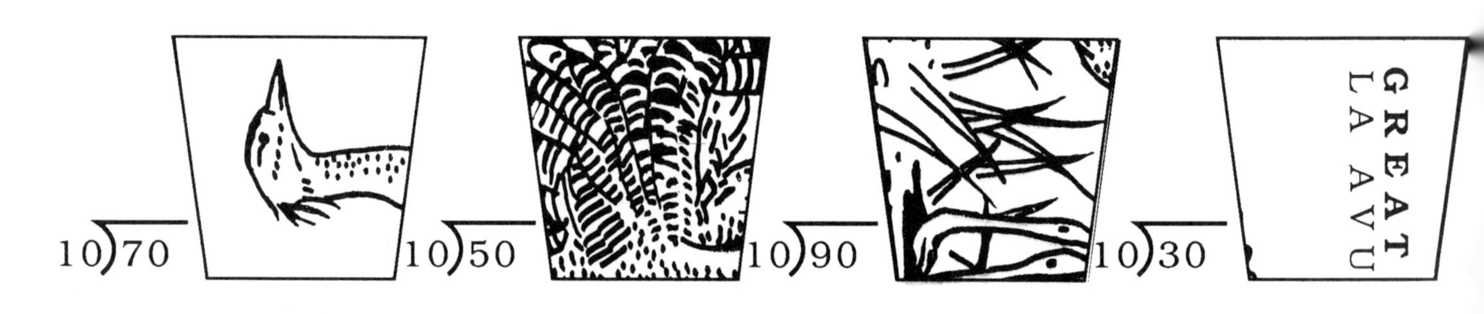

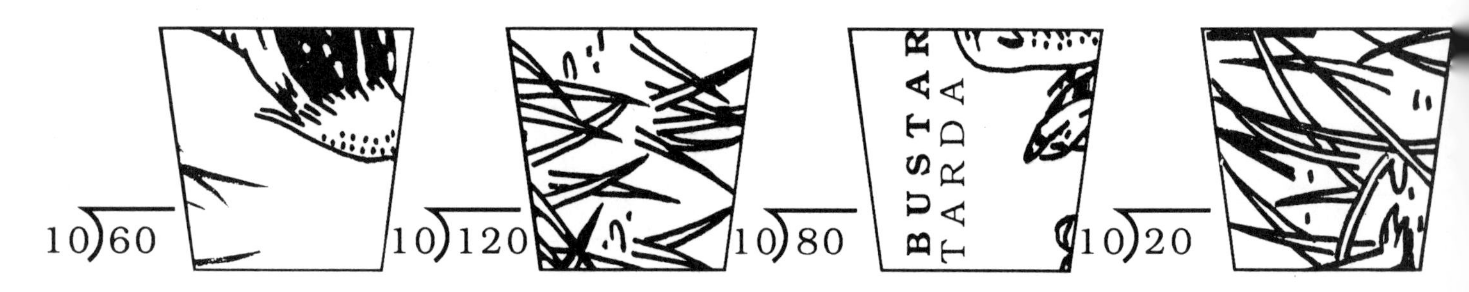

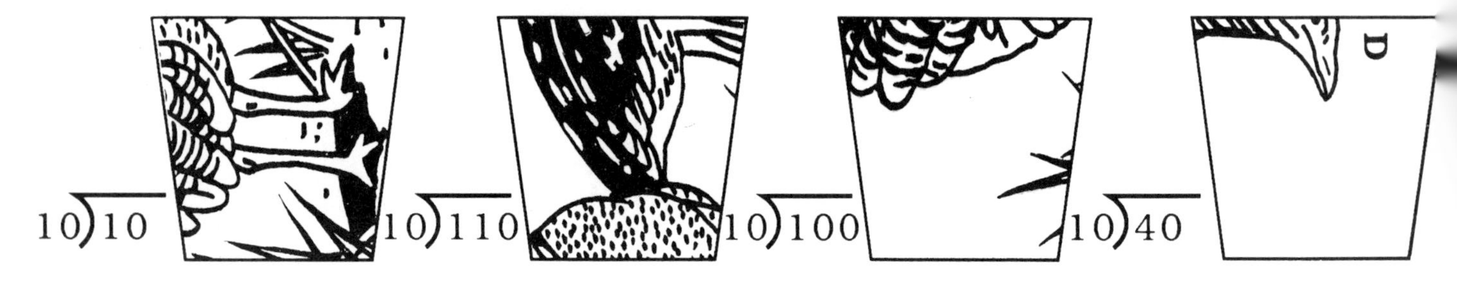

NAME
NOMBRE ______________________________

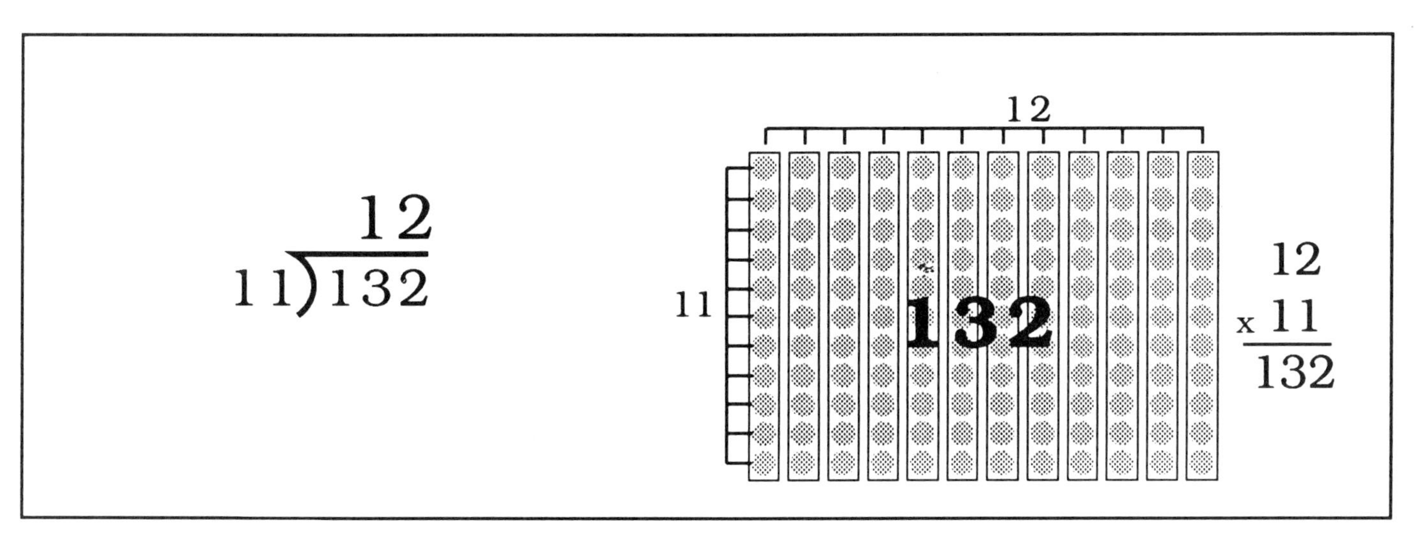

$11\overline{)11}$ $11\overline{)99}$ $11\overline{)33}$ $11\overline{)0}$ $11\overline{)132}$ $11\overline{)55}$ $11\overline{)22}$ $11\overline{)110}$

$11\overline{)66}$ $11\overline{)44}$ $11\overline{)121}$ $11\overline{)88}$ $11\overline{)11}$ $11\overline{)33}$ $11\overline{)77}$ $11\overline{)99}$

$11\overline{)110}$ $11\overline{)132}$ $11\overline{)55}$ $11\overline{)77}$ $11\overline{)22}$ $11\overline{)121}$ $11\overline{)33}$ $11\overline{)66}$

$11\overline{)88}$ $11\overline{)22}$ $11\overline{)132}$ $11\overline{)0}$ $11\overline{)99}$ $11\overline{)110}$ $11\overline{)44}$ $11\overline{)121}$

NAME
NOMBRE ______________________________

Instructions:

1. **Answer all the math problems first.**
2. **Cut out one puzzle piece at a time.**
3. **Paste the puzzle piece in the box with the same answer.**

Instrucciones:

1. Conteste todos los problemas de matemáticas primero.
2. Recorte una pieza del rompecabezas a la vez.
3. Pegue la pieza del rompecabezas en el recuadro que tiene la misma respuesta.

6	**10**	**1**	**5**
4	**12**	**3**	**9**
7	**2**	**8**	**11**

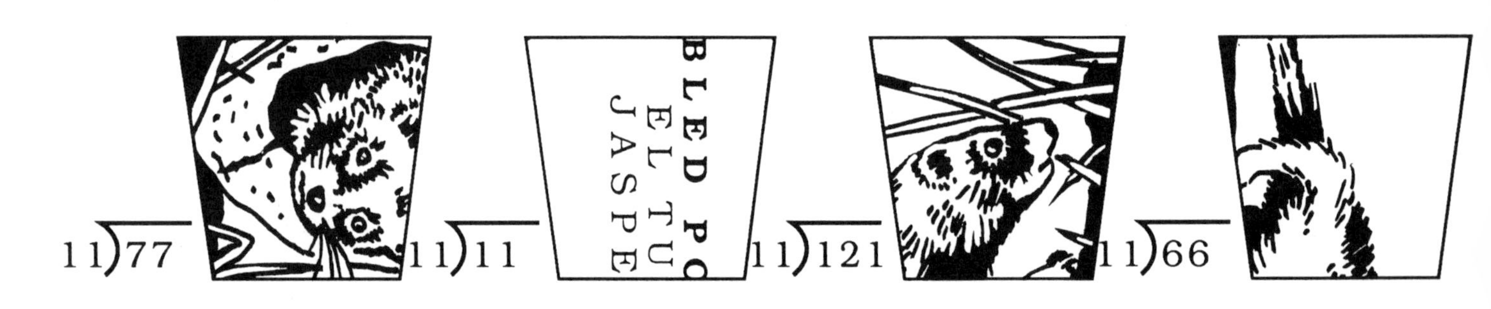

$11\overline{)77}$ $11\overline{)11}$ $11\overline{)121}$ $11\overline{)66}$

$11\overline{)33}$ $11\overline{)22}$ $11\overline{)55}$ $11\overline{)110}$

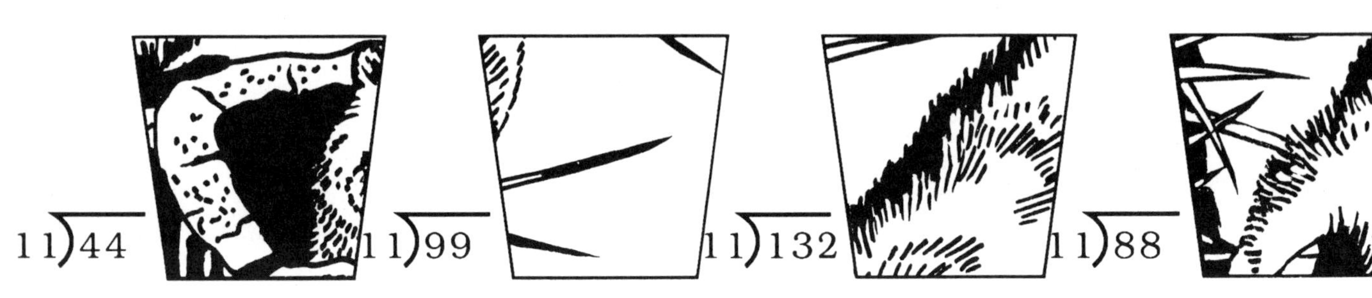

$11\overline{)44}$ $11\overline{)99}$ $11\overline{)132}$ $11\overline{)88}$

NAME
NOMBRE ______________________________

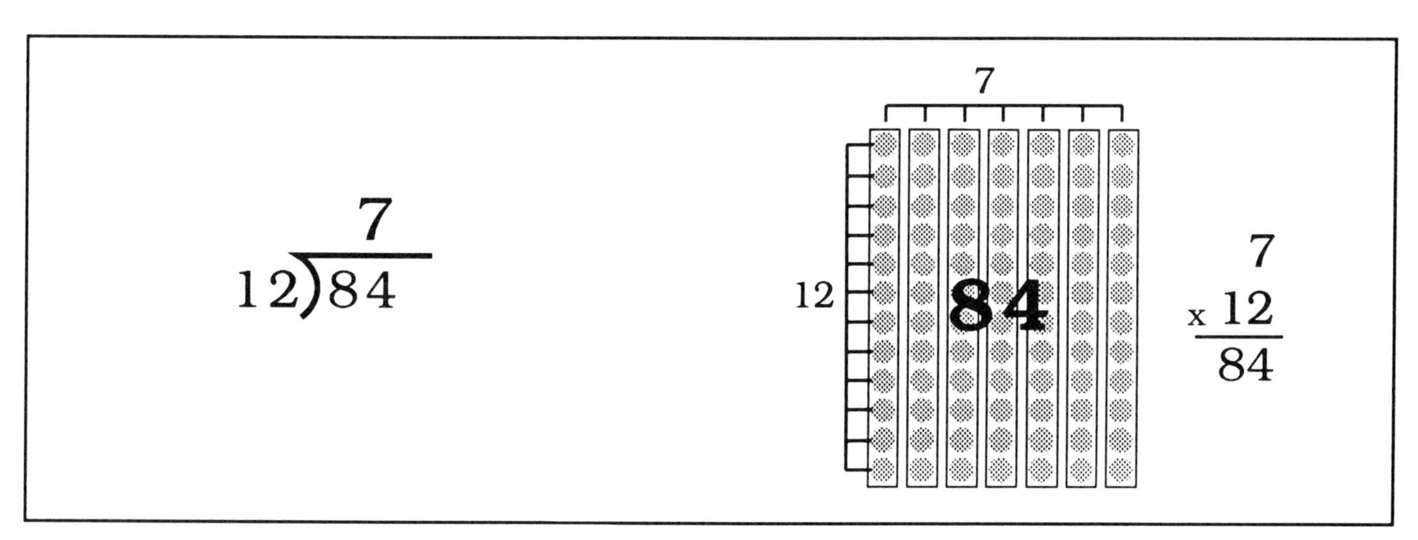

12)24 12)144 12)48 12)0 12)108 12)84 12)60 12)132

12)96 12)12 12)120 12)60 12)36 12)144 12)72 12)0

12)108 12)24 12)48 12)120 12)84 12)132 12)12 12)96

12)144 12)36 12)60 12)12 12)132 12)96 12)24 12)72

Lesson 12
Lección 12

NAME
NOMBRE ________________________________

Instructions:

1. **Answer <u>all</u> the math problems first.**
2. **Cut out <u>one</u> puzzle piece at a time.**
3. **Paste the puzzle piece in the box with the same answer.**

Instrucciones:

1. Conteste <u>todos</u> los problemas de matemáticas primero.
2. Recorte <u>una</u> pieza del rompecabezas a la vez.
3. Pegue la pieza del rompecabezas en el recuadro que tiene la misma respuesta.

1	**3**	**5**	**7**
10	**12**	**9**	**11**
2	**4**	**6**	**8**

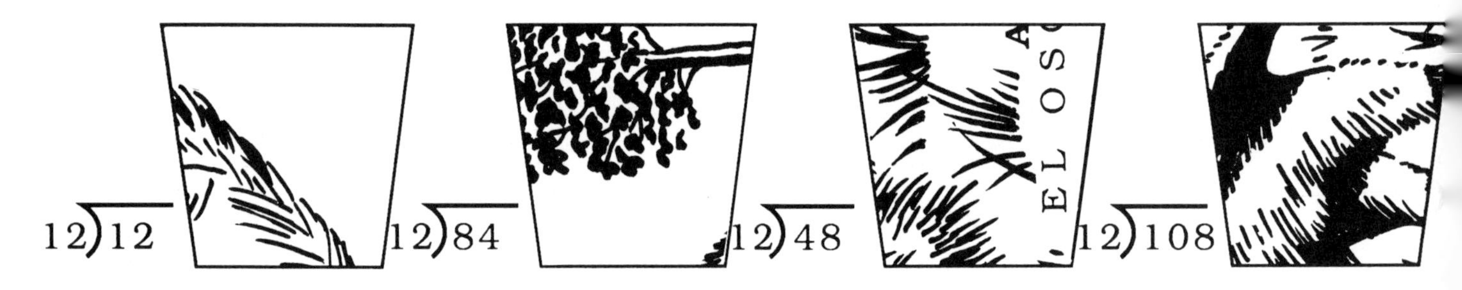

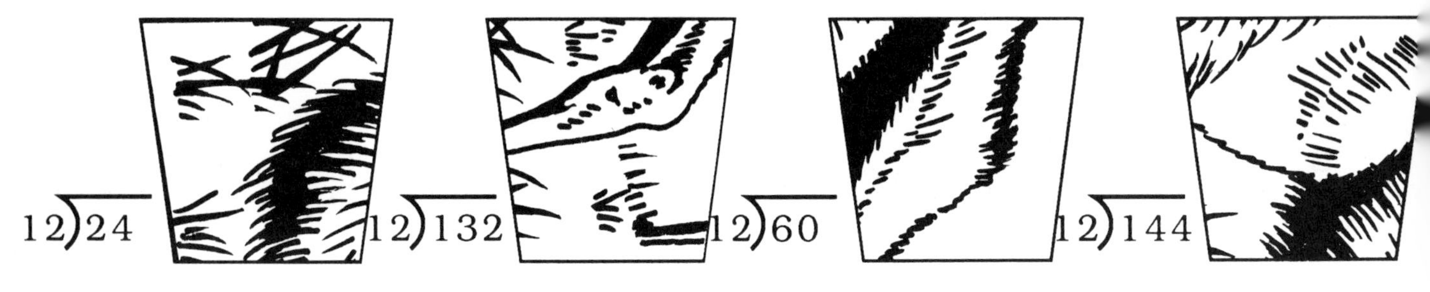

Division Post-Test

NAME
NOMBRE ______________________________

$9\overline{)36}$	○ 5 / ● 4 / ○ 6
$3\overline{)18}$	○ 5 / ○ 7 / ○ 6
$6\overline{)24}$	○ 4 / ○ 5 / ○ 3
$9\overline{)108}$	○ 11 / ○ 10 / ○ 12
$12\overline{)96}$	○ 9 / ○ 8 / ○ 7
$2\overline{)6}$	○ 3 / ○ 4 / ○ 2
$5\overline{)35}$	○ 5 / ○ 7 / ○ 6
$8\overline{)40}$	○ 7 / ○ 6 / ○ 5
$1\overline{)121}$	○ 12 / ○ 10 / ○ 11
$1\overline{)9}$	○ 9 / ○ 8 / ○ 1
$4\overline{)48}$	○ 11 / ○ 9 / ○ 12
$7\overline{)21}$	○ 4 / ○ 3 / ○ 2
$0\overline{)100}$	○ 10 / ○ 1 / ○ 11
$12\overline{)60}$	○ 6 / ○ 5 / ○ 7
$8\overline{)72}$	○ 9 / ○ 6 / ○ 8
$11\overline{)132}$	○ 11 / ○ 9 / ○ 12
$9\overline{)63}$	○ 14 / ○ 12 / ○ 7
$11\overline{)110}$	○ 11 / ○ 12 / ○ 10
$7\overline{)56}$	○ 6 / ○ 8 / ○ 9
$12\overline{)144}$	○ 12 / ○ 11 / ○ 10

Answers

Lessons 1 -12 & Post-Test

Page 1.

4	7	10	2	11	0	1	8
10	3	5	0	9	12	6	9
2	5	8	6	12	1	4	10
3	9	11	4	7	12	1	8

Page 3.

10	7	1	11	3	6	12	2
4	2	8	5	1	9	5	3
2	4	12	0	9	11	3	10
6	8	11	4	7	1	10	0

Page 5.

3	2	8	6	0	12	9	4
9	5	1	7	10	1	5	11
12	2	6	0	4	7	11	6
8	10	3	12	7	1	2	5

Page 7.

9	3	0	12	2	6	11	4
1	10	7	4	10	7	8	5
2	8	5	12	4	3	11	0
12	1	6	11	9	7	3	8

Page 9.

12	5	9	0	7	4	8	3
9	6	2	10	1	5	11	2
6	1	12	0	3	11	5	10
8	3	10	9	4	12	7	1

Page 11.

4	1	8	12	3	0	5	9
5	8	2	7	11	4	7	1
10	6	8	11	1	3	9	12
3	10	12	9	4	11	6	0

Page 13.

9	3	12	0	10	7	5	2
4	6	11	3	1	12	8	4
6	2	12	8	5	7	10	0
7	4	9	11	2	6	3	1

Page 15.

6	0	2	8	12	4	10	1
3	7	11	0	5	9	2	12
1	5	12	8	4	7	10	3
7	11	9	1	8	3	6	10

Page 17.

6	3	12	8	10	0	5	2
9	7	1	6	4	8	10	3
8	4	12	2	5	0	3	11
7	9	11	1	6	12	2	7

Page 19.

1	12	5	0	9	2	4	10
3	11	9	1	5	12	7	3
2	10	8	5	12	0	6	9
4	6	10	8	11	3	1	7

Page 21.

1	9	3	0	12	5	2	10
6	4	11	8	1	3	7	9
10	12	5	7	2	11	3	6
8	2	12	0	9	10	4	11

Page 23.

2	12	4	0	9	7	5	11
8	1	10	5	3	12	6	0
9	2	4	10	7	11	1	8
12	3	5	1	11	8	2	6

Page 25 - Post-Test.

4	6	4	12
8	3	7	5
11	9	12	3
10	5	9	12
7	10	8	12

Giraffe

The giraffe is found in Africa. It is the tallest animal and feeds on leaves and twigs that are 20 feet (6m) above the ground. The giraffe's keen sight, smell, and hearing protect it from many dangers.

La Jirafa

La jirafa se encuentra en Africa. Es el animal de más estatura y se alimenta de hojas y ramitas que se encuentran hasta 20 pies (6m) de altura. Los buenos sentidos de vista, olfato, y oido la protejen de muchos peligros.

Cheetah

The cheetah is found in Africa. Its slim body and long legs enable it to run up to 75 miles per hour (121 km) for short distances, It is a skillful hunter that takes advantage of its speed to catch its prey.

El Leopardo Cazador

El leopardo cazador se encuentra en Africa. El cuerpo esbelto y las piernas largas le permiten correr hasta 75 millas por hora (121km) por cortos trechos. Es un cazador hábil que se aprovecha de la velocidad para alcanzar su presa.

Ostrich

The ostrich is found in Africa. It is the largest living bird. It reaches 8 feet (2.4m) in height and can weigh 350 lbs. (160 kg). The ostrich cannot fly, but its powerful legs enable it to run up to 30 miles per hour (48km).

El Avestruz

El avestruz se encuentra en Africa. Es la mayor de las aves actuales. Alcanza los 8 pies (2.4m) de altura y puede llegar a pesar 350 libras (160kg). El avestruz no puede volar, pero las piernas poderosas le permiten correr hasta 30 millas por hora (48km).

Warthog

The warthog is found in Africa. It is about 5 feet (1.5m) long and travels in family groups. Its upper canine teeth curve up over its snout, making it a fierce-looking animal.

El Jabalí Verrugoso

El jabalí verrugoso se encuentra en Africa. Mide unos 5 pies (1.5m) de largo y se mueve en grupos familiares. Los colmillos superiores se encorvan hacia arriba encima del hocico, dándole un aspecto feroz.

Lion		El León
The lion is found in Africa. Lions live in family groups. Lions hunt mostly at night, stalking and surprising their prey.		El león se encuentra en Africa. Los leones viven en grupos familiares. Por la mayor parte, los leones cazan de noche, acechando y sorprendiendo a su presa.
Armadillo		**El Armadillo**
The armadillo is found in North and South America It hunts at night. It uses its powerful front feet to burrow and to dig for insects, worms, and small creatures. When it is threatened by a larger animal, it rolls into a ball and plays dead.		El armadillo se encuentra en Norteamérica y Sudamérica. Caza de noche. Utiliza las poderosas patas delanteras para hacer sus madrigueras y para desenterrar insectos, gusanos y bichos pequeños. Cuando es amenazado,se enrolla en forma de pelota y se hace el muerto.
Vulture		**El Buitre**
The vulture is found throughout the world. It is a meat eater and preys on small animals and dead carcasses. A vulture has good senses of sight and smell.		El buitre se encuentra por todo el mundo. Es un carnicero que apresa y devora pequeños animales y come carne podrida. El buitre tiene buenos sentidos de vista y de olfato.
Bison		**El Bisonte**
The bison is found in North America. Bison travel in large herds. They migrate south in the fall to avoid harsh winters. In the spring, a herd moves north again, grazing on fresh grasses.		El bisonte se encuentra en Norteamérica. Los bisontes se mueven en manadas grandes. Emigran al sur en el otoño para escaparse de los inviernos severos, y en la primavera las manadas se trasladan de nuevo al norte, pastando en las nuevas hierbas.